In Commemoration of the 75th Anniversary of the Crash of the Airship USS Shenandoah 1925 – 2000

OHIO'S AIRSHIP DISASTER

By the author of

"Black Hand Gorge...A Journey Through Time"
An illustrated history of a spectacular river gorge in Licking County, Ohio.

and

"Statues on the Hill"
The remarkable story of an undertaker and the forgotten statues he carved on an isolated hill in Muskingum County, Ohio.

Ohio's Airship Disaster

The Story of the Crash of the USS Shenandoah

by

AARON J. KEIRNS

Little River
PUBLISHING

HOWARD, OHIO

*To Bernice,
Adam, Tracy, Jesse and Nathan*

♥

Additional copies of this book can be ordered from:

Little River Publishing
P.O. Box 291
Howard, OH 43028

Phone: 740-599-5189

ISBN 0-9647800-2-X

Little River
PUBLISHING

©2000 by Little River Publishing
P.O. Box 291 • Howard, Ohio 43028

ALL RIGHTS RESERVED

The information printed herein was obtained from a variety of sources including old books, newspapers, photographs, documents and personal interviews. We have attempted to verify all information but, due to the historical nature of the subject matter, accuracy is not always certain. Due to the age of many of the photographs, the photographer or copyright-holder could not be determined. The publisher welcomes comments from readers.

Book design by Aaron Keirns

Shenandoah Monument in Ava, Ohio

"To the memory of those who lost their lives on the U.S.S. Shenandoah, September 3, 1925"

Commander Zachary Lansdowne, U.S.N.
Lieutenant Commander Lewis Hancock, Jr., U.S.N
Lieutenant John Bullard Lawrence, U.S.N
Lieutenant Arthur Reginald Houghton, U.S.N.
Lieutenant (JG) Edgar William Sheppard, U.S.N
Everett Price Allen, ACR, U.S.N
Charles Harrison Broom, ACMM, U.S.N.
James William Cullinan, AP, U.S.N.
Ralph Thomas Joffray, AR1c, U.S.N.
Celestino Pellegrino Mazzuco, AMM1c, U.S.N.
James Albert Moore, Jr., AMM1c, U.S.N.
Bartholomew Bennett O'Sullivan, AMM1c, U.S.N.
George Conrad Schnitzer, CRM, U.S.N.
William Howard Spratley, AMM1c, U.S.N.

Contents

Foreword – *by C. Lansdowne Hunt* 6

Preface – *by Aaron J. Keirns* 7

Chapter 1
The Dreams and Dangers of Flight 9

Chapter 2
The Golden Age of Airships 13

Chapter 3
The Daughter of the Stars 19

Chapter 4
The Shenandoah's Final Flight 29

Chapter 5
The Aftermath 37

Chapter 6
The Legacy of the Shenandoah 53

Addendum 60
Glossary of Terms 60
Acknowledgements 61
About the Author 61
Bibliography 62

Ohio's Airship Disaster   5

Foreword

When the author offered me the privilege of writing this foreword, I was delighted to see, upon reviewing his manuscript, that he had employed a reader-friendly writing technique. Aaron Keirns has given a "fresh face" to the wealth of fact and lore surrounding one of the Twentieth Century's principal aviation events.

Reaching back to the oral tradition passed down from my grandmother, Lieutenant Commander Lansdowne's widow, I can recall that she said that the family dog, Barney, began barking for no apparent reason in the early morning hours of September 3, 1925. According to my mother, Peggy, who was a month away from her third birthday, the family could not calm the English bull terrier occupying their quarters at the Lakehurst Naval Air Station. Within hours, Betsy Ross Lansdowne would learn that she was a 23-year-old widow.

This thoroughly enjoyable work gives excellent perspective to our nation's experimentation in Lighter-than-Air (LTA) by devoting the early chapters to a factual, but not laborious, treatment of the airship's evolution in Europe. My mother used to say that her father had the best possible career path for an American acquiring an education in airships: "The Germans knew how to build 'em and the Brits knew how to fly 'em." Lieutenant Commander Lansdowne's career before SHENANDOAH included a tour at the LTA facility at RAF Howden, England and as Assistant Naval Attache to Commander "Bull" Halsey at the U.S. Embassy in Berlin, Germany.

In the later chapters devoted to the short life of the U.S. Navy's first rigid, Mr. Keirns skillfully focuses on the actions of the various personalities in the air and on the ground as they intertwine that fateful Wednesday morning. Liberally sprinkling the text with photos and drawings taken at and around the various crash locations, the author creates for the reader the illusion of actually witnessing this terrible event. There are fascinating and little known factual accounts which are not even chronicled in my family archives. These accounts, enhanced by the author's skill at telling the story, are of special interest and enjoyment to those who have an interest in Ohio history and those of us descended from the individuals involved. Ample proof of Mr. Keirns' excellent research is evidenced by his judicious use of sidebars disclosing human interest stories and reconstructing fundamental events associated with the crash.

Reflecting on the 75th Anniversary of SHENANDOAH's crash, I am confident that you will enjoy this volume as much as I have, in that it is a quality work that coincides with the public's rekindled interest in the country's brief experiment with rigid airships at the dawn of Naval Aviation.

C. Lansdowne Hunt
September 3, 2000

Preface

There's an old house standing on a hill near Ava, Ohio, with a story to tell. The wind blows freely through its missing windows, and tall pine trees cast ominous shadows across its weathered clapboards.

It's here that a farmer named Andy Gamary once lived with his wife and five children. One morning in early September of 1925, something incredible fell out of the sky into Andy's yard and garden. With a series of ground-shaking thuds, the USS Shenandoah, America's first rigid airship, came tumbling out of a turbulent cloud and slammed into Ohio history.

The crash of the Shenandoah was a major event at the time. It was front page news in newspapers all across America. But as the years have passed, the dramatic loss of this pioneering airship has all but faded from our collective memory.

It's probably safe to say that most Ohioans are more familiar with the crash of the Hindenburg (which happened in New Jersey) than with the crash of the Shenandoah. Most would also be surprised to learn that the Shenandoah crashed several years before the Hindenburg even existed.

The Shenandoah was truly an important part of aviation history. She was America's first rigid dirigible – sometimes referred to as the "Model-T" of airships. She was also the first rigid airship in the world to be filled with helium instead of explosive hydrogen. Her captain, Lieutenant Commander Zachary Lansdowne, was one of the world's foremost airship experts, a pioneer in the field of lighter-than-air flight. He was also a native Ohioan.

A visit to the Old Gamary house is a moving experience. Standing alone on its quiet hill, there is no evidence of the calamity and chaos that once fell upon this peaceful place. But there is a sense of history here. It's something you can feel but not express. Much like the feeling you get when walking among the monuments at Gettysburg. You can feel that something dramatic happened on this spot in another time.

Who knows, maybe someday there will be a permanent museum or perhaps an historic park at the site of the Shenandoah crash. A place where the story can be told with pictures, displays and walking tours of the crash sites. A place to learn, reflect and keep alive the memory of the great airship and her crew. Time will tell.

I hope you enjoy this book. It's not intended to be the "last word" on the subject. There have been other books and articles written about the Shenandoah which have probed much deeper into the subject. But unfortunately most of these resources are now out-of-print.

The purpose of this book is to bring together, in one place, a simple story of the life and death of the airship Shenandoah. I've included a lot of pictures – they often tell the story better than words. I hope this book will help re-acquaint Ohioans, and other interested folks, with this unique and important event in Ohio history.

Aaron J. Keirns
September 3, 2000

8   *Ohio's Airship Disaster*

CHAPTER 1

THE DREAMS AND DANGERS OF FLIGHT

The dream of breaking free from our earthly bonds and soaring across an open sky is as old as mankind itself.

The ancient Greek myth of Icarus tells of human-powered flight on wings of waxed feathers. During the Middle Ages and Renaissance, men envisioned complicated flying contraptions with mechanical wings that flapped like a bird's. Artist and visionary Leonardo da Vinci even sketched an idea for the helicopter as early as the year 1483.

At one time or another, earth-bound humans have conceived of all sorts of strange methods and devices to enable themselves to reach the heavens – from the European idea of attaching birds to a harness, to the Chinese plan of tying rockets to a kite. Some of these ideas were pure fantasy, others were just ahead of their time.

LIGHTER-THAN-AIR FLIGHT

One idea that eventually caught on was the notion of lighter-than-air flight. This was a simple idea. A vehicle would rise into the air not by complex mechanics, but simply because of its ability to float. Much like a ship floats on the water, a lighter-than-air "ship" would be buoyant enough to float on the air. Not surprisingly, some of the early designs of lighter-than-air craft were very ship-like in appearance. Even the word "gondola," used to describe the basket or car suspended below a balloon or airship, refers to a type of flat-bottomed boat such as those used on the canals of Venice.

THE RISE OF HOT AIR

It had long been known that a lightweight container would rise if filled with hot air. It was only a matter of time before someone thought of using this principle to carry a man aloft. In fact, some researchers theorize that hot-air balloon flight may have actually been accomplished by early cultures, long before our recorded history. On the plains of Nazca in southern Peru are huge depictions of animals and other shapes outlined

Above: A balloon ascension in Paris in 1784. Notice the ship-like shape of the gondola carrying the passengers.
(Musée de l'Air, Paris)

Opposite page: Icarus falls from the sky as the sun melts his wings.
(National Air and Space Museum Smithsonian Institution)

Right: This curious contraption was named the Passarola (Great Bird). During its test flight in 1709 it could only flap helplessly on the ground.
(National Air and Space Museum, Smithsonian Institution)

Above: Madeleine Blanchard falls to her death when her hydrogen-filled balloon explodes over Paris in 1819, becoming the first woman to die in an air accident.
(Musée de l'Air, Paris)

J. A. C. Charles

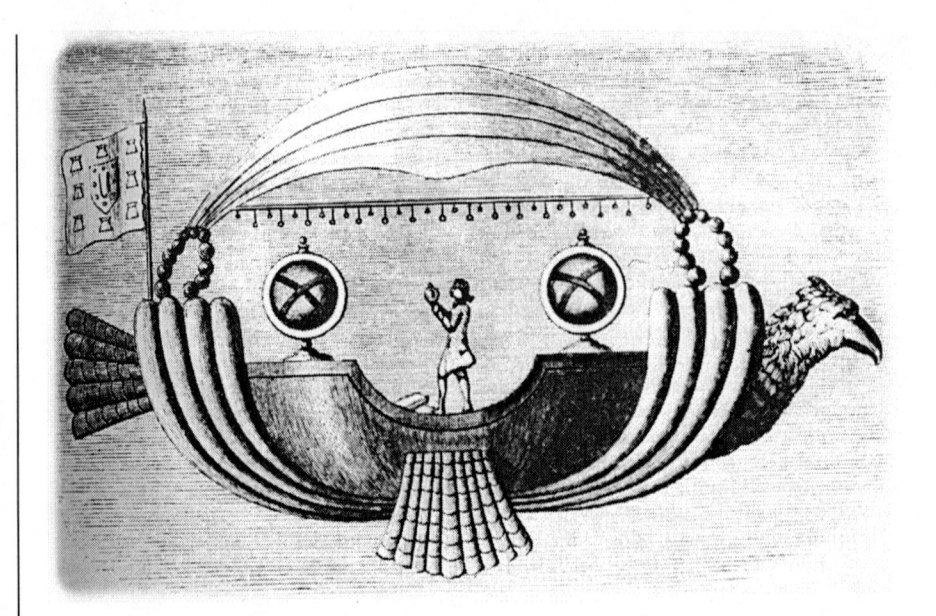

with stones by early man. The stylized shapes are indiscernible from the ground, but strikingly apparent when viewed from the air. Could these early people have accomplished flight? If so, it may have been in some kind of hot-air balloon.

The inherent problem with hot-air balloon travel however is that the ability of hot air to supply lift decreases as the air cools. Unless a controllable heat source can be carried aloft with the balloon, hot-air flights are relatively short and hard to control. Today's hot-air balloons are equipped with sophisticated burner systems that enable balloons to travel great distances. Modern balloonists have the ability to employ bursts of heat at will, giving them a fair amount of control. But this kind of technology is a recent innovation. For many years lighter-than-air flight was possible but not very practical. Finally, in the late 18th century, a Frenchman came up with a better idea.

IT'S A GAS

An alternative to using hot air was conceived by a French chemist named J. A. C. Charles. In 1783 Charles successfully ascended in a balloon filled with hydrogen gas. Hydrogen is the lightest of all known elements. It is naturally buoyant and doesn't need to be hot to rise. This was a great advantage over using heated air. After this initial success, hydrogen would be used in all kinds of balloons and airships for well over a century.

THE EVOLUTION OF THE AIRSHIP

After Charles' flight in 1783, many other successes (and failures) followed. One of the biggest problems with these early attempts was a lack of control. The balloonists were essentially riding a free-floating bubble of gas (or hot air) with little control over direction or speed. Landings were little more than planned accidents, often depositing pilots on roof tops or in trees.

Left: Santos-Dumont's semi-rigid Airship No. 5 near Paris in 1901.
(National Air and Space Museum, Smithsonian Institution)

Every small success and failure added to the experience and knowledge of these early aeronauts. As time passed, new ideas and innovations lead to more practical approaches.

Controlled Flight

In 1898 Alberto Santos-Dumont, a wealthy Brazilian living in Paris, became the first man to mount an engine on an airship, marking the beginning of controlled flight. His first ship, appropriately named "Airship No. 1," was powered by a 3.5 horsepower engine borrowed from his three-wheeled motorcycle. Between 1898 and 1905 Santos-Dumont built a total of 14 hydrogen-filled, semi-rigid airships of various designs, becoming the toast of France and Brazil in the process.

About the same time, but on the other side of the Atlantic, something quite different was going on. Two brothers from Ohio, Orville and Wilbur Wright, were pioneering heavier-than-air flight. In 1903, Orville climbed aboard the Wright brothers' flying machine, named the *Flyer*, flew for 12 seconds and traveled a distance of about 120 feet along the sand at Kitty Hawk, North Carolina. The Flyer was powered by a 12 horsepower engine which drove two propellers. The Wrights were not well-known and, at first, the world just yawned at the news of their short flights. But soon the significance of their achievements would become recognized worldwide.

The success of the Wright brothers and others who were working on heavier-than-air vehicles, threatened to cut short the development of lighter-than-air craft – at least for a while. It was apparent that heavier-than-air craft held a greater potential for speed and maneuverability than the huge lumbering airships. But airship enthusiasts held on to their dreams. They believed the future of flight held a place for both airplanes and airships. The skies were certainly big enough for both, and for several years airplanes and airships developed and matured alongside one another.

Thus by 1903 after centuries of trial & error, both lighter-than-air and heavier-than-air vehicles had achieved some initial success. The window to the skies was finally beginning to open.

An Early Wright Biplane, ca. 1904.
(National Air and Space Museum, Smithsonian Institution)

Sky Monster

In the 1780's two French brothers launched a 15 ft. hydrogen-filled balloon into the rainy skies above Paris. After drifting 15 miles, it came down to earth near the little village of Gonesse.

The local peasants were terrified at the sight of this strange apparition descending upon them, and proceeded to hack it to shreds with scythes and pitchforks.

12 *Ohio's Airship Disaster*

Chapter 2

The Golden Age of Airships

In the first third of the 20th century, the idea of lighter-than-air flight reached a peak of popularity. There was a certain fascination and romance associated with airships. There were airship-themed songs, poems, postcards, stamps, toys and children's books. Airships became an important part of our popular culture.

Airships held great promise as passenger transports, cargo haulers and even bombers. From roughly 1903 to 1937 the airship industry grew and prospered. During this period there were great success and terrible failures. It was an exciting time in the history of flight – a time when giants roamed the skies.

Balloons, Blimps, Dirigibles and Zeppelins

Balloons, blimps, dirigibles and zeppelins are all words that bring to mind lighter-than-air craft. But these words each have specific meanings.

Balloons and *blimps* are terms that refer to non-rigid airships. The word blimp is a combination of "b" for balloon and "limp" which is how it looks before inflation. In a balloon or blimp the envelope containing the hot-air or gas has no rigid internal structure. These airships essentially take shape only when inflated. The famous Goodyear® blimp is an example of a non-rigid ship.

The word *dirigible* literally means "capable of being steered." Thus any steerable rigid, non-rigid or semi-rigid airship could fall into this broad category.

Zeppelin is a term used to describe rigid airships. The word is taken from the name of the most famous designer of rigid airships, Count Ferdinand von Zeppelin of Germany. Zeppelins have an internal structure, usually consisting of a complex metal framework, that define their shape. The lifting gas is contained in individual "cells" or bags located within the internal framework. The Shenandoah was the first rigid airship

An elaborate die-cut airship card.

Opposite page: In the early 1900's, the airship craze coincided with a postcard craze. The airship became a popular theme for all occasions. Postage for postcards was 1 cent.

Right: In this old postcard, a view of the "near future" included a variety of fantasy airship-type vehicles.

Above: A whimsical postcard from the early 1900's. Like other technical innovations, the airship became a symbol of man's progress.

Even jewelry designers were inspired by the airship motif.

to be built in the United States.

There is also a type of airship known as *semi-rigid*. These ships employ elements of both non-rigid and rigid craft. A semi-rigid ship is fundamentally a balloon or blimp with a rigid keel attached *(see photo on page 11)*. Many early airships were of this type. Interestingly, the Zeppelin Company of Germany has recently begun production on a series of new semi-rigid airships which may lead the way for a renewed interest in airship production.

HYDROGEN & HELIUM

The rigid airships of the early 1900's typically used hydrogen gas to provide lift. Hydrogen, the most abundant element in the universe, was easily accessible and economical. The big problem with hydrogen was its flammability and explosiveness. Any airship filled with hydrogen was a floating bomb. Several hydrogen-filled ships suffered fiery crashes as a result of utilizing this volatile gas.

In the late 1800's a new lighter-than-air, non-flammable, non-explosive gas was discovered: helium. Although much safer than hydrogen, helium had its own problems – it was rare, very expensive and only available in the United States.

When the Shenandoah was built by the United States Navy in 1923, it became the first rigid airship to utilize helium. But because helium was so expensive and hard to get, hydrogen-filled ships remained the norm well into the late 1930's.

THE AIRSHIP AT WAR

From the very beginning, the value of lighter-than-air vehicles for use in war was recognized by military men. Observation balloons were used as early as 1794 by the French in the Battle of Fleurus. Balloons were also employed by troops during the Civil War in the United States during the 1860's.

Airships took a much more active role during World War 1, however. When war broke out between Great Britain and Germany, the airship came of age as a weapon. Beginning in 1915, rigid airships fitted with machine guns and bombs brought terror and death to British towns and cities. Always attacking at night, the German airships were difficult to detect from the ground. They could also rise above the range of British aircraft. In the end however, actual death and destruction caused by the German airship attacks was slight (by World War standards). Probably the most important effects of the German airship attacks were the public terror they inflicted and their ability to wear down the British and tie up their troops and planes.

By World War II the strategies and machinery of war had changed. Rigid airships didn't play much of a role this time. Non-rigid ships did, however. The United States Navy used more than 150 blimps during World War II. These blimps served mostly as antisubmarine patrols and convoy escorts. No vessel under escort from a Navy blimp was ever sunk by an enemy sub.

THE GREAT ZEPPELINS

In order to understand the significance of the airship Shenandoah, it's helpful to look at some of the other well-known ships that came before and after her.

Probably the most tireless experimenter in the history of modern airships was Count Ferdinand von Zeppelin of Germany. The Count was an ex-cavalry commander who was determined to take the design of rigid airships to new heights. His persistence helped bring the rigid airship to prominence during the 1920's and 1930's. Even today, many people still refer to any airship as a "zeppelin."

Count Zeppelin didn't originate the idea of a rigid airship, but he was a pioneer in making it work. Zeppelin's rigid framework was made of an exotic new metal called aluminum. It was both light and strong. A tough fabric was then stretched over the framework. The lifting gas (which in Zeppelin's case was hydrogen) was contained in several cells or bags located within the rigid structure. This basic design would become the standard for later airships, including the USS Shenandoah.

Zeppelin's first rigid airship, the LZ-1, flew in 1900. He later formed a company and went on to create several more advanced ships, one of which was the famous *Graf Zeppelin* designated as LZ-127. Launched in 1928, the "Graf," named after the Count himself, (Graf is the German word for Count) was the largest dirigible built to that date. It was 776 feet long and was powered by five 530-horsepower engines.

In October of 1928 the Graf flew across the Atlantic, carrying a crew of 40, plus 20 passengers. The trip took 111 hours and covered 6,168 miles. The trip was a great success and led the way for future transatlantic passenger flights. This ship went on to provide a regular transatlantic passenger service for several years. It eventually crossed the ocean 144 times

Count Zeppelin.

Above: Passengers could dine in style aboard airships like the Graf.
(Library of Congress)

Postage stamps featuring the Graf.

The Graf Zeppelin.
(Library of Congress)

and carried over 40,000 people. The Graf proved that rigid airships could be a reliable and safe way to transport people over long distances.

But in 1937 something happened that brought an abrupt end to airship passenger service. The German airship *Hindenburg* exploded into flames while attempting to land at Lakehurst, New Jersey.

THE HINDENBURG DISASTER

The Hindenburg, or LZ-129, was built with lessons learned from the success of the Graf Zeppelin. Hindenburg was even bigger than the Graf. At 804 feet long it was 28 feet longer and had nearly twice the gas capacity. Hindenburg could carry 72 passengers and had a range of 8,000 miles.

The Hindenburg had originally been designed to be filled with helium, but the United States wouldn't release the gas to the Germans. There was a fear that Germany might use the helium for military purposes. Because of this embargo, the Hindenburg was inflated with hydrogen instead. This unfortunate circumstance lead to its untimely demise.

Right: The Hindenburg crashing in flames at Lakehurst, New Jersey.

16 *Ohio's Airship Disaster*

On May 6, 1937, the Hindenburg floated in for a routine landing at Lakehurst, New Jersey. At 7:25 pm, to the horror of all who watched, the great ship suddenly exploded into a giant fireball, killing 36 of the 97 passengers and crew on board. The huge ship was reduced to ashes in a matter of seconds. As flaming pieces of the airship came crashing down, spectators on the ground ran for their lives. A news crew captured the chaotic scene on film.

This singular disaster became etched into the collective consciousness of the world by the dramatic film footage shot that day in Lakehurst. Other fatal dirigible crashes had come and gone (including the Shenandoah) but none had caught the attention of the world quite like the Hindenburg disaster.

The crash of the Hindenburg brought an abrupt end to the golden age of great airships. There had been too many crashes and too many deaths over the years. The graphic news reels and photographs of the terrible Hindenburg crash had a chilling effect on the public's perception of airship travel in general. Even the well-proven Graf Zeppelin was quickly taken out of service and dismantled after the Hindenburg went down. The dramatic crash of the Hindenburg effectively slammed the window shut on the era of the great passenger airship.

To Conquer the Sky

It took centuries of trial and error for man to realize his dream of flight. But even with today's sophisticated flying machines, man is never truly at home in the air. His ability to defy gravity is temporary at best. Every flight is still subject to mechanical failure, human error and the weather. The entire history of man's attempt to conquer the skies is riddled with tragic stories of his chronic inability to stay aloft.

Airships have certainly contributed their share of tragic stories to the history of flight. Often, unpredictable weather has been the cause. Big, slow-moving airships are easy targets for the wrath of mother nature. Over the years, winds and storms have tossed and twisted many of these helpless giants, sending them back to earth in pieces. The USS Shenandoah was one of these unfortunate ships. She rose triumphantly into the skies only to be cast down in a crumpled heap.

Like Icarus who plummeted to Earth when his waxen wings melted by flying too close to the sun, the USS Shenandoah, christened the *Daughter of the Stars*, met a similar fate when she flew too close to a storm.

One September morning in 1925 the Daughter of the Stars slipped and fell from the sky, crashing into Ohio history.

This is her story…

What Caused the Hindenburg to Explode?

It has long been assumed that leaking hydrogen, perhaps from a punctured gas cell, caused the fiery explosion of the Hindenburg.

Addison Bain, a retired NASA engineer offers a different explanation. Bain, who is an expert in hydrogen, has spent several years studying the incident and the materials involved. He believes that the initial fire may have started on the outer covering of the dirigible, not by the ignition of free hydrogen.

In a 1997 article, Bain said the materials and coatings used on the outer fabric of the Hindenburg were extremely flammable. The cotton base fabric (itself flammable) was coated with a combination of cellulose acetate or nitrate and an aluminum powder. As Bain points out, nitrate is used to make gunpowder and aluminum powder is used as fuel in today's solid rocket boosters.

Bain believes the coated fabric of the Hindenburg's skin may have been ignited by an electrical spark which only then ignited the hydrogen inside.

©1997 Popular Science Magazine

18   *Ohio's Airship Disaster*

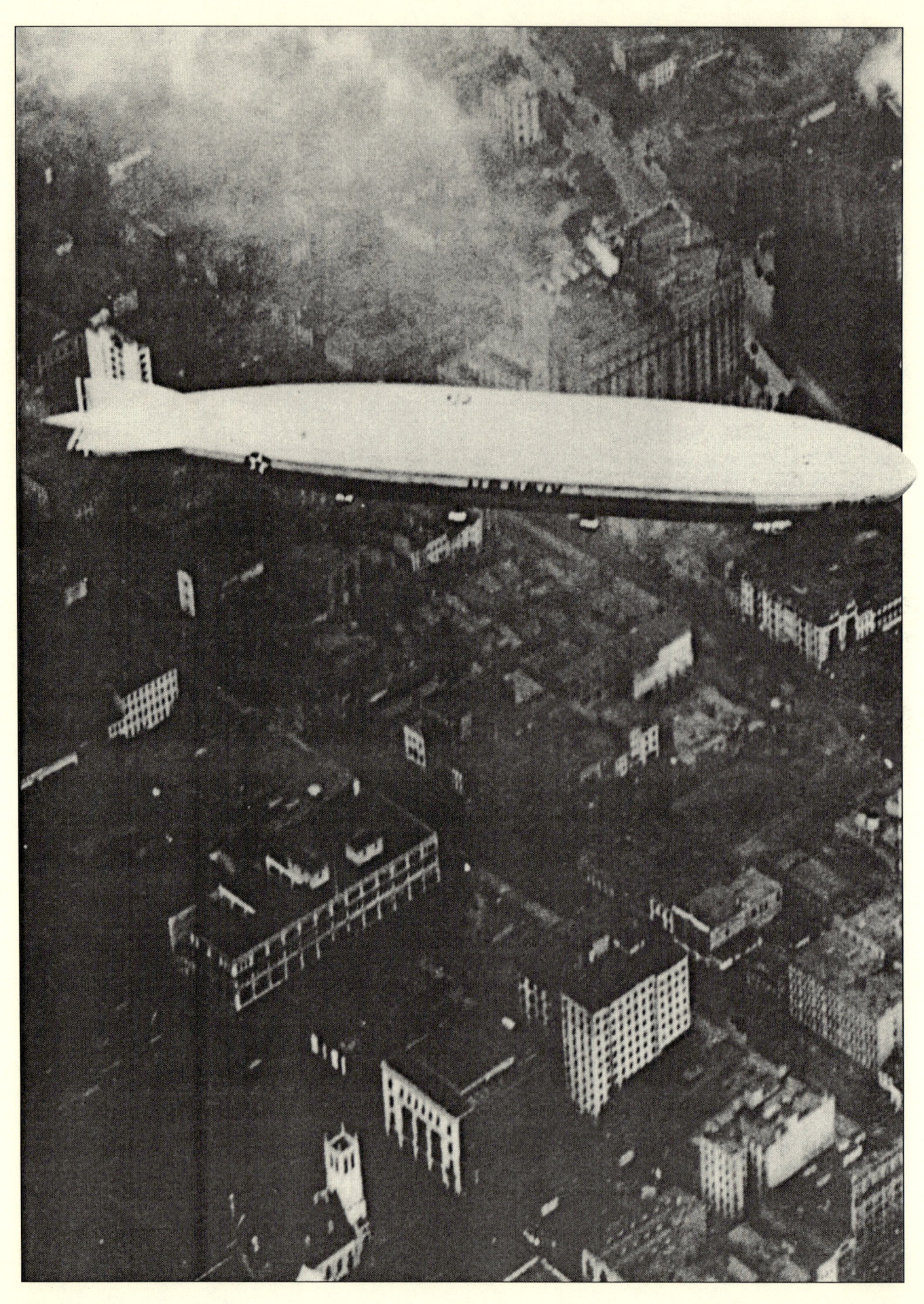

CHAPTER 3

THE DAUGHTER OF THE STARS

It was a time of great enthusiasm for airships. In 1919 Great Britain's rigid airship R-34 left Scotland bound for the United States. It carried 30 passengers and crew. The R-34 battled wind and storms much of the way but managed to reach land after 108 hours and 12 minutes aloft – a new record.

By the time the ship reached its destination on Long Island a huge crowd of enthusiastic Americans had converged on the spot. A detail of 500 military policemen were sent to provide crowd control. More than 1,000 men from the U.S. Navy Air Service were also on hand to assist with the landing. But no one on the ground was experienced in landing a large rigid ship. So an officer on the ship, Major Pritchard, strapped on a parachute and jumped. He landed with a jolt and was immediately surrounded by reporters. "Can you tell us what your first impressions of America are, sir?" asked one reporter. "Hard," was Pritchard's reply.

Upon reaching America, Brigadier General Edward Maitland, the captain of the R-34, looked down from his ship and wrote in his logbook, "…what a small place this world really is…and what an astonishing part these great Airship Liners will play in linking together the remotest places of the earth; and what interesting years lie immediately ahead!" General Maitland was a great advocate of airship travel. Little did he know that in the years "immediately ahead," an airship would carry him to his death.

AMERICA CLIMBS ABOARD

General Maitland's enthusiasm over the achievement of the R-34 was shared by people on both sides of the Atlantic. Unlike the airplanes of the time, rigid airships could carry passengers in style and comfort and travel much greater distances. The future of airship travel seemed bright.

America didn't yet have an airship of its own, but that was about to change. The same year that Britain's R-34 touched down in America, the U.S. Congress appropriated funds for a Navy airship base and the acqui-

Above: A young naval officer named Zachary Lansdowne was aboard the R-34 as an observer during its flight from Scotland to the United States in 1919. A few years later he would become the celebrated captain of America's first rigid airship, the Shenandoah.

Opposite page: The Shenandoah over Lakehurst, New Jersey in 1924. *(U. S. Air Force)*

Ohio's Airship Disaster

Right: A stereograph showing the British airship R-34. Stereographs contained two separate images taken at slightly different angles. When viewed with a stereoscope viewer, they produced a 3-D effect. Viewing stereographs was a popular pastime before the advent of the television.

sition of two rigid airships. One of these airships, the ZR-1 (later christened the Shenandoah) was to be built in the United States. The other ship, the R-38, was already under construction by the British and would be re-designated as ZR-2. (All of the U.S. Navy's lighter-than-air ships were given the designation "Z" for zeppelin-type and "R" for rigid).

Disaster Strikes – Twice

The British R-38 never made it to the United States. In 1921 while making a final test flight, it broke in two during maneuvers over England. The forward half of the ship burst into flames and fell into the Humber River while the rear half landed on a sandbar. A total of 16 Americans and 28 Britons were killed in the crash. Among the dead was General Maitland.

Even after the R-38 disaster, America's enthusiasm for the airship remained strong. At the Naval Aircraft Factory in Philadelphia work began on the ZR-1 and the Zeppelin Company of Germany was hired to build a replacement for the R-38.

Right: Wreckage of the British R-38 (ZR-2) in the Humber River.
(U. S. Navy)

Then, in February of 1922, it happened again. An Italian-made, semi-rigid airship named the "Roma," which had been purchased by the United States Army was being test flown from Langley Field, Virginia. About 45 minutes into the flight the Roma suddenly lost altitude, striking high-tension wires. The Roma's hydrogen gas exploded and the ensuing crash killed 34 people.

The fiery crashes of the R-38 and Roma had one positive side effect. They convinced the Americans that airships should be filled with a non-flammable gas. Inert helium was the gas of choice. It was heavier than hydrogen but it was non-flammable. The safety of helium came at a high price however. It cost $55 or more to produce 1,000 cubic feet of helium as opposed to about $3 for the same amount of hydrogen. Fortunately for the United States, the only place in the world where helium was known to exist was in the gas fields of Texas and Kansas.

DEBUT OF THE SHENANDOAH

The Navy's ZR-1 (the Shenandoah) would become the first rigid airship to be filled with non-flammable helium. The design of the ZR-1 was based on the captured German airship L-49. While participating in a raid on England in 1917 the L-49 had been forced down in France by a storm and was captured intact, allowing engineers to study its construction.

Above: To help put the size of the Shenandoah into perspective, these two silhouettes show the size difference between a well-known Goodyear® blimp (on top) and the long, slender Shenandoah.

The Goodyear blimp measures 205 ½ feet long. The Shenandoah was 682 feet long – that's longer than 2 football fields.

The USS Shenandoah.
(Library of Congress)

Although the hydrogen-filled L-49 served as a model for the Navy's ZR-1, it was not copied exactly. Since the ZR-1 would use helium as its lifting gas, certain modifications had to be made. Helium has less lift than hydrogen, so a 30 ft. section was added to the middle of the ZR-1 to allow for more gas capacity. This extension gave the ZR-1 a long, graceful appearance.

On September 3, 1923, fifteen thousand spectators showed up at the Naval Air Station in Lakehurst, New Jersey. They were there to witness the maiden flight of the ZR-1, America's first rigid airship. The flight lasted about an hour, boosting even higher the American public's enthusiasm for airships.

Mrs. Edwin Denby, wife of the Secretary of the Navy, was given the honor of naming the ZR-1. Mrs. Denby christened the great ship "Shenandoah" in honor of her beloved home in the beautiful Shenandoah Valley of Virginia. The name Shenandoah is believed to be a native American word meaning "Daughter of the Stars," a fitting name for a graceful airship.

Shenandoah Specs

Length: 682 ft.

Diameter: 78 ft. 9 in.

Total Weight: 41 tons

Gas Cells: 20

Engines: Five 300 h.p. Packards (originally it had six engines)

Cruising Speed: 60 mph

Right: The Shenandoah moored to her mast at Lakehurst, New Jersey. *(U. S. Navy)*

Above: Advertisers were quick to capitalize on the popularity of the Shenandoah, as these two full-page magazine ads demonstrate.

THE NORTH POLE FLIGHT

Riding the wave of airship mania, President Coolidge approved a plan to fly the Shenandoah on an exploratory trip to the North Pole. The trip was to take place in the Spring of 1924. Airplanes of the time lacked the necessary range for polar flights and the crew of the Shenandoah expected to be the first to reach the Pole by air. It would be a great adventure.

Then on December 21, 1923, yet another great airship met a fiery demise. This time it was the French "Dixmude," an airship with a record of impressive feats. Dixmude had once taken a long flight over the Sahara Desert setting a new world record for endurance in the air. But now, the Dixmude, returning to France from a long flight, had exploded over the Mediterranean Sea during a violent thunderstorm. All 50 men on board were killed. It was the worst air disaster yet.

The Americans were saddened by this news but their enthusiasm was little dampened. After all, this was just another disaster caused by the explosiveness of hydrogen. The Shenandoah, with its non-flammable helium, would not have this problem. Preparations for the Shenandoah's polar flight continued.

An Unplanned Adventure

In January, 1924, the Shenandoah was moored to its mast at Lakehurst to undergo a ten-day wind test. On her flight to the North Pole she would need to be able to withstand stiff Arctic winds. For four days the tethered ship battled gusts of wind occasionally reaching 60 miles per hour. The Shenandoah performed well. Then, on January 16, a winter storm hit Lakehurst. Wind gusts increased until the strain was too much, ripping the Shenandoah from her moorings. As the nose of the ship headed downward, crewmen immediately released 4000 pounds of water ballast. The ship began to rise. Then to lighten the ship even more, crewmen in the keel released three huge gasoline tanks which exploded when they hit the ground.

The crippled ship drifted north in the storm while the captain and crew tried to regain control. In the stormy dark they passed over Newark, then on to New York City. Emergency repairs were made by the crew and, after several frightening hours, the winds finally began to subside. Nine hours after her ordeal began, the Shenandoah limped back home to

Above: On her first trip to Washington, D.C., the Shenandoah flew over the Capitol, then dropped flowers on the grave of the Unknown Soldier in Arlington.
(National Geographic Magazine, 1925)

Below: The Shenandoah moored to the USS Patoka. The Shenandoah was the first airship to be moored to a floating mast.
(U. S. Navy)

Lakehurst, damaged but not defeated.

This sobering incident caused a postponement of the Shenandoah's planned trip to the North Pole.

ZACHARY LANSDOWNE

There had been some dissension in the ranks aboard the Shenandoah and the Navy decided it was time to bring in a new captain for the ship. In February of 1924, Lieutenant Commander Zachary Lansdowne reported to Lakehurst to take over command.

Lansdowne was born in Greenville, Ohio, northwest of Dayton, a town also known for being the hometown of famous sharpshooter Annie Oakley. Lansdowne was a tall man, 6'-1", with high cheekbones and the distinguished bearing of an officer.

He was reportedly a stern but affable officer and well-respected by his men. One of the United States' most experienced airship officers, he had been on board the R-34 on its transatlantic flight in 1919 and later had tested the feasibility of helium as an alternative to hydrogen aboard a Navy blimp. His presence brought about fresh enthusiasm for the planned North Pole excursion.

Lansdowne was also working on a manual for the operation of large rigid airships. In the manual he made several recommendations for making airship flight safer. Ironically, one recommendation he made was warning against flying near thunderstorms. He cautioned that their wrenching updrafts and downdrafts could be disastrous for any airship regardless whether it was filled with hydrogen or helium. This observation would prove all too true in years to come.

THE TRANSCONTINENTAL FLIGHT

The damage caused by the Shenandoah's unplanned trip to New York City was quickly repaired and the airship was again made ready for duty. Over the next few months several successful flights were made with Commander Lansdowne in command. On August 8, 1924, off the coast of Rhode Island, the Shenandoah successfully moored to the tanker ship, *Patoka*, becoming the first rigid airship to be moored to a floating mast. News of this success again boosted enthusiasm for the North Pole flight.

The next test for the Shenandoah would be a cross-country flight. The trip would allow the captain and crew to test several mooring masts that had been erected at strategic locations across the United States in preparation for her North Pole excursion. The trip would also provide a great public relations opportunity for the U.S. Navy.

On October 7, 1924, the Shenandoah began her long cross-country journey. Junius B. Wood, a reporter for The National Geographic Magazine had been given the honor of covering the trip as a passenger. National Geographic devoted 47 pages of its January 1925 issue to the trip. Wood described the scene as the Shenandoah was led out of its hangar at Lakehurst: "Every man on the station helped, 300 of them –

Zachary Lansdowne

Above: Early 1920's writing tablet featured the Shenandoah on its cover.

Opposite page: Magazine ad from 1924 featuring the Shenandoah.
(The American Magazine)

SHENANDOAH -- INDIAN WORD MEANING "DAUGHTER OF THE STARS"

When the "Daughter of the Stars" talks with the children of earth

YOU remember the dramatic night last winter when the giant Navy dirigible Shenandoah went adrift in a raging gale.

"You are over Newark," said radio station WOR. "What can we do to help you?"

Thousands of people sitting by radio sets in their cozy homes heard the plucky lieutenant-commander on the Shenandoah send back the reply: "Thanks, old man, everything's O. K."

In the air, as on the sea, radio equipment must be the most reliable it is possible to get. That is why the Shenandoah, the huge ship Leviathan —in fact, many government and commercial radio plants—were equipped with Exide Batteries.

For your own set

When *you* use Exide Radio Batteries in your home you get the clearest reception, for Exides give uniform current through a long period of discharge.

The new Exide 6 volt "A" Battery in one-piece case.

There is an Exide type for every tube and a size for every set: "A" batteries for 2-volt, 4-volt and 6-volt tubes; "B" batteries, 24 and 48 volt, of 6000 milliampere hour capacity. They are efficient, dependable, long-lasting — and right in price. Exide Batteries are made by the largest manufacturer of storage batteries in the world.

You can get Exides at Radio Dealers and at all Exide Service Stations. If your dealer should be out of booklets describing Exide Radio Batteries, send us your name and we will mail them to you.

THE ELECTRIC STORAGE
BATTERY COMPANY
Philadelphia

Exide Batteries of Canada, Limited
153 Dufferin St., Toronto

Exide
RADIO BATTERIES

FOR BETTER RADIO RECEPTION USE STORAGE BATTERIES

sailors, marines, Filipino mess boys, and civilians. They came running into the drear, misty morn like little ants pulling an immense gray worm out of its nest."

The trip would cover more than 9,000 miles. Beginning at Lakehurst, New Jersey, the Shenandoah headed south through the southern states and into Texas. Then along the Mexican border and up the coast all the way to Seattle, Washington. On its return trip, the ship followed a similar path back as far as Fort Worth, Texas, then turned north up through the midwest. The Shenandoah entered Ohio at Greenville, and passed over the boyhood home of Commander Lansdowne. Far below in her yard, stood the commander's mother, Elizabeth, waving as the ship and her son went floating by.

From Greenville the ship's route dipped slightly to Dayton then back up to Springfield and over Columbus. It then continued due east, passing over Zanesville, on course for Moundsville, West Virginia. As the Shenandoah flew east from Zanesville it passed just north of Ava, only a few miles from the spot where, less than a year later, it would come crashing to the ground.

During the trip across the continent, the Shenandoah had maneuvered its way out of several tight spots, including almost hitting a mountain while crossing the Rockies. But the trip was considered a great success and the airship had proven itself to a nation. The Shenandoah had become the pride of Americans from coast to coast.

THE SISTER SHIP ZR-3

While the Shenandoah was still on her transcontinental flight, America's new German-made airship ZR-3 arrived in Lakehurst from Germany. The ZR-3 was the airship the U.S. Navy had ordered from the Zeppelin Company in order to replace the ZR-2 (British R-38) which had crashed before delivery in 1921.

The ZR-3, christened the *Los Angeles*, came across the Atlantic using hydrogen, but the hydrogen was removed after it reached Lakehurst. Like the Shenandoah, the Los Angeles would be filled with helium. The problem was that there was not enough helium to go around. Humorist Will Rogers reportedly quipped that the Navy now had "…two airships but only one helium."

The only solution was to borrow helium from the Shenandoah. So the Shenandoah's helium was transferred into the Los Angeles. For several months the Shenandoah sat gasless while the Los Angeles used its helium to make several successful trips.

Then the Los Angeles ran into difficulties during a flight to Minneapolis. The ship had to turn back to Lakehurst where it was discovered that some major repairs were needed. Finally, it was time to revive the Shenandoah. In late June, 1925, the precious helium was bled out of the Los Angeles and back into the gas cells of the lifeless Shenandoah.

The Los Angeles (ZR-3). Note how the control car is mounted directly to the main body of the ship.

A Change of Plans

During the time the Shenandoah sat idle in the hangar at Lakehurst, plans for the North Pole trip were scrapped. President Coolidge had reconsidered and decided the trip would be too risky and expensive. Instead, the Shenandoah would make demonstration flights around the American Midwest visiting several state fairs. The trip was scheduled for late August.

As a native Ohioan, Zachary Lansdowne knew from experience how volatile the weather could be in the midwest during late summer. It has been written in some accounts that he was not at all pleased with the idea of flying the airship around the midwest in this stormy season. Trying to adhere to a tight schedule of fair appearances added another element of danger to the trip.

But according to C. E. Rosendahl, navigator aboard the Shenandoah during its final flight, Commander Lansdowne was not overly concerned about the possibility of bad weather during this trip. "I shall never forget my asking Captain Lansdowne as we were about to cast off from the Lakehurst mooring mast, what kind of weather map we had for our midwest flight, and his reply that the map was a 'good one' and that 'yesterday's map was even better'," said Rosendahl.

Rosendahl had worked closely with Lansdowne and was adamant that the experienced captain would not have made the trip against his better judgement, no matter who demanded it.

"To assume that through blind obedience, an officer of Zachary Lansdowne's caliber and future would knowingly place his ship and his crew in serious jeopardy of their lives is wholly inconsistent with the facts," Rosendahl said in his remarks at the 50th anniversary of the crash held in Ava, Ohio in 1975. (Rosendahl's remarks were actually presented in his absence by author/historian Lewis Gray).

In any case, Lansdowne did officially request a postponement of the trip. He wanted to wait at least until the second week of September in order to assure that all the various landing fields would be ready. His superiors denied the request because some of the fairs on the schedule would be over by then.

In the end the trip was scheduled for the first week of September. No one could have known at the time just how perilous this decision would prove to be for the Shenandoah and her crew.

Commander Lansdowne in the control car of the Shenandoah.
(Brockway Collection)

A Case of Semantics

When the midwest trip was being planned, Commander Lansdowne wrote to the Bureau of Aeronautics that the midwest trip should be made "...after the thunderstorm season is over (September)."

Author, Thom Hook in his 1973 book "Shenandoah Saga," suggests that the Bureau of Aeronautics didn't interpret the month in parenthesis as a month in question but rather as a safe month in which to make the trip.

As Hook points out, the sentence could just as easily be interpreted to mean that the thunderstorm season is in September, and the trip should be made <u>after</u> that month.

Columbus Evening Dispatch

OHIO'S GREATEST HOME DAILY

THE ONLY AFTERNOON PAPER IN COLUMBUS RECEIVING THE ASSOCIATED PRESS DISPATCHES

FULL PAGE OF PICTURES — Interesting photographs of events and persons are contained in a full page feature in The Dispatch daily.

COLUMBUS, OHIO, THURSDAY, SEPTEMBER 3, 1925.   VOL. LV. NO. 65.   PRICE TWO CENTS.

WEATHER—Fair tonight. Friday fair and cooler.

SHENANDOAH, TWISTED BY SQUALL, PLUNGES DOWN IN THREE PIECES

FOURTEEN MEMBERS OF CREW INCLUDING LT. COM. LANSDOWNE KILLED NEAR CALDWELL, OHIO

Most of Dead are Found in Wreckage of Control Cabin Which Dropped in Cornfield on Outskirts of Ava

TAIL SECTION DRIFTS 12 MILES

Out of Control in High Wind, Big Dirigible Bumps Ground Several Times and Then Seals Fate by Crashing into Walnut Tree

THE DEAD.

Lieutenant Commander Zachary Lansdowne, Greenville, Ohio, commanding officer.
Lieutenant Commander Louis Hancock, Austin, Tex., second in command.
Lieut. J. B. Lawrence, St. Paul, Minn., senior watch officer.
Lieut. A. R. Houghton, Allston, Mass., watch officer.
Everette P. Allen, Omaha, Neb., aviation chief machinist.
Charles Broom, Toms River, N. J., aviation chief rigger.
... men W. Cullinan, Binghamton, N. Y., aviation pilot.
... T. Joffray, St. Louis, Mo., aviation rigger.
... celestine P. Mazzuco, Murray Hill, N. J., aviation machinist mate.
James A. Moore, Jr., Savannah, Ga., aviation machinist mate.
Bartholomew O'Sullivan, Lowell, Mass. aviation machinist mate.
George C. Schnitzer, Tuckerton, N. J., chief radio man.
William H. Spratley, St. Louis, Mo., aviation machinist mate, first class.
Lieut. E. W. Sheppard, Washington, D. C., engineer officer.

THE INJURED.

Raymond Cole, Lima, Ohio, radio chief.
John F. McCarthy, Boston, Mass., aviation chief rigger.

CALDWELL, OHIO, SEPT. 3.—(AP)—The giant dirigible Shenandoah is no more. It went down in three pieces here early today and killed its commander, Lieutenant Commander Zachary Lansdowne and at ... making up her crew.

SHENANDOAH AS IT APPEARED BEFORE CRASH

COMMANDER ZACHARY LANSDOWNE

EYE-WITNESS TELLS HOW HE SAW AIR MONSTER FALTER, BREAK, AND SLIP TO EARTH

C. L. Arthur Declares Tragedy Was Preceded by Muffled Roar; Gondola Carrying 13 Dashed to Ground Like Egg-Shell

BELLE VALLEY, OHIO, SEPT. 3. ...

PASSENGER HURTLED OUT.

SHENANDOAH TRAGEDY STORY IS ONE OF HEROISM OF CREW WHO FOUGHT STORM IN VAIN

Col. C. G. Hall Tells of Thrilling Battle Put Up by Men Who Dangle on Ropes and Ladders High in the Air

AIR CURRENTS TOO STRONG

Blames Failure of Meteorological Stations to Warn Dirigible of Dangerous Conditions of Atmosphere

CALDWELL, OHIO, SEPT. 3.—The story of the Shenandoah disaster, in which the giant dirigible crashed and broke into three portions over Ava, Ohio, near here, this morning, resulting in the death of 14 aboard the airship, is one of heroism of crew, pioneers in the interest of the development of lighter-than-air transportation. It is best told by Col. C. G. Hall, U.S. army observer aboard the ill-fated ship.

"We were traveling west at an altitude of about 3000 feet when we encountered a storm," Col Hall said in describing the accident. "By changing our course a dozen or more times, we dodged it, only to encounter ...

Navy Department Hears That Lightning Wrecked Big Dirigible

WASHINGTON, SEPT. 3.—(AP)—The Moundsville, W. Va., aviation field telegraphed the navy department today that the Shenandoah was "struck by lightning" at 6:35 o'clock this morning.

The message indicated the information had been obtained from army aviators who had gone to the scene.

"Shenandoah struck by lightning," the message said, "during storm at 5:35 a. m. today near Pleasant City, Ohio, south of Cambridge. Ship cut in half. One part down at Pleasant City. Other part down at Berne, about two miles east of Caldwell, Ohio. Positions verified by Major Kerr, air service, flying from Fairfield to Langin this a. m."

...the line squall which sent us to an altitude of 5500 feet ...realized what had happened.

THE CINCINNATI ENQUIRER

CITY EDITION

VOL. LXXXII. NO. 247—DAILY   FRIDAY MORNING, SEPTEMBER 4, 1925   TWENTY PAGES   PRICE FIVE CENTS

Shenandoah Is Destroyed; Fourteen Are Killed; Wilbur, Warned, Insisted on Flight, Is Charge

POLITICS!

Is Cry of Widow.

Lansdowne Fought Trip, Knowing Ohio Perils,

But Superior Desired Publicity, 'Tis Said.

Had He Known Outcome! Is Ironic Comment.

Girl Takes Chance on Flier, Is Philosophy.

I've Gambled and Lost! Sorrowing One Says.

GRAPHIC PHOTOS OF ILL-FATED SHENANDOAH

El Paso in Grip of Severe Flood;

LIST OF CASUALTIES

Craft Caught in Ohio Gale

Front Cabin Is Death Trap For Commander and 12 Others.

Score Ride Ten Miles on Wreckage and Descend Without Injury.

Third Section Alights in Grove With Three Clinging To Torn Rigging.

"Line Squall" Brings End To Pride of Navy—Greenville Officer Perishes With Crew—

---

28   Ohio's Airship Disaster

CHAPTER 4

THE SHENANDOAH'S FINAL FLIGHT

On September 2, 1925, the Shenandoah left Lakehurst bound for several midwest cities including Columbus, St. Louis, Minneapolis and Detroit. It was to be a six-day trip covering about 3,000 miles. The Navy considered it a perfect opportunity to show off its famous airship at fairs across the midwest.

As the ship lifted off, the weather reports appeared favorable. Soon the airship passed over Philadelphia then floated on across Pennsylvania. During the night she slipped over the Alleghenies. As the ship crossed into Ohio everything seemed fine and the captain was asleep in his quarters. Then, around 3:00 a.m., the situation began to change rapidly.

INTO THE STORM

As the ship neared Cambridge, Ohio, flashes of lightning began to appear off to the northwest. The wind was picking up and the sky began to look threatening. The ship was struggling to make any headway against the wind. Commander Lansdowne was called back to his post. An ominous black storm cloud lay dead ahead. Lansdowne ordered the ship to turn south hoping to go around the worst of the danger. The severity of the storm was not immediately obvious, but another huge, wind-swept cloud was forming just above the great ship. It was almost as if two storms had conspired to converge directly upon the Shenandoah. The airship was caught in a deadly line squall, strong winds were pushing her long slender hull upward and downward at the same time. The turbulence was getting worse by the minute and strong headwinds were now bringing all forward progress to a standstill. The captain and crew stood by their controls doing everything possible to maneuver the ship out of harm's way.

A little before 5:00 a.m., the ship began rising uncontrollably, faster and faster, briefly leveling off at 3,000 feet – then it shot up again at a rate of 1,000 feet per minute. Helium was quickly released and, at about 6,200 feet, the ascent was brought to a halt. Then the ship went into freefall. A lot of helium had been vented and the ship began to plummet toward the

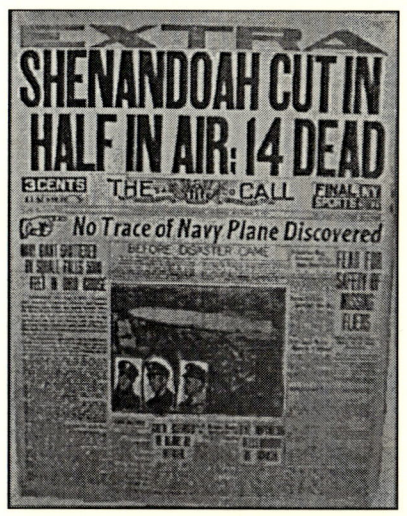

Above: All across the country, special editions of newspapers called "extras" carried the story of the tragic crash.

Opposite page: Front pages of the Columbus Evening Dispatch, and the Cincinnati Enquirer.

Ohio's Airship Disaster 29

The Line Squall

The storm that tore apart the Shenandoah has often been characterized as a "line squall".

A line squall is a strong sudden wind (or squall) that occurs along a line of thunderstorms. Line squalls are usually short-lived but can be very violent.

Shortly after the crash of the Shenandoah, the United States Weather Bureau gave their interpretation of what probably took place during what they referred to as the "Shenandoah Squall."

"Caught in the updraft of the storm, the ship was driven aloft…and then broke to pieces, presumably under the strain of winds blowing upward and downward at the same time – or at any rate with very different vertical velocities – on different parts of her long hull."

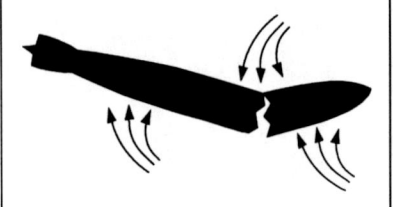

The article continued by saying: "…an ordinary balloon or aeroplane would not have been exposed to a similar strain and would probably have weathered the gale."

Charles Fitzhugh Talman
United States Weather Bureau
From an article in Nature Magazine November, 1925, entitled "Who's Who Among the Storms"

Opposite page: This diagram shows a highly simplified version of the sequence of events that took place during the crash of the Shenandoah.

---

earth at a rate of 1,500 feet per minute, even faster than it had risen.

Crewmen immediately released over 4,000 pounds of water ballast to slow the descent. It seemed to be working. The ship began to slow down and finally leveled off at around 3,000 feet. As rattled crewmen breathed a short sigh of relief a rapidly rising air mass enveloped the ship again, pushing it back skyward at an alarming rate of speed. The long, slender Shenandoah began to spin and convulse under the tremendous strain. Now the nose of the ship was pointing almost straight up while gale-force winds wrenched and twisted the ship like two giant hands twisting apart a loaf of French bread.

THE BREAKUP & CRASH

Internal support cables began to snap and the ship's metal framework gave out an eerie metallic cry as it twisted violently and began pulling apart. The strain was so intense that the ship cracked open like a giant egg, spilling two helpless crewmen out into the dark sky – the first two casualties of the storm. Finally the ship ripped completely in two with only a few control cables holding the huge sections together.

Up front in the control car Commander Lansdowne told his men they could leave the car if they wanted to. Two of the men quickly scrambled up the ladder into the forward hull, the rest, including Lansdowne, stayed at their posts.

The control car, suspended below the ship, was beginning to shake violently now and the struts holding it to the hull were starting to pull away. Commander Lansdowne and his men held on for their lives. Then in one final thrust the control car broke free from the ship. As it fell it yanked loose the control cables that ran through the keel back to the rudder and elevators. The long cables ripped violently through the ship as the car and its occupants plummeted toward the earth. Seconds later the car smashed into the ground killing all on board. Engines #4 and #5, along with their crewmen, also came crashing down pulling much of the midsection down with them.

MEANWHILE, DOWN ON THE GROUND…

Andy Gamary, a Noble County farmer, was awakened around 5:45 a.m. by some jarring thumps and scraping noises near his house. He cautiously ventured outside to discover an astounding and grisly sight. Huge mounds of twisted metal and silvery fabric now occupied the yard around his home. Scattered among the masses of tangled debris lay the contorted bodies of several men. There were twelve in all. All dead except one who died a few minutes later.

Andy's wife, Mary, and their children came out and were stunned by the sight. They had no idea who these men were or what this huge thing was that had fallen out of the sky into their yard and garden. It appeared to be some sort of aircraft, but they weren't sure what it was. The Gamarys hadn't heard about the Shenandoah's midwest trip, in fact, they had never even heard of the Shenandoah.

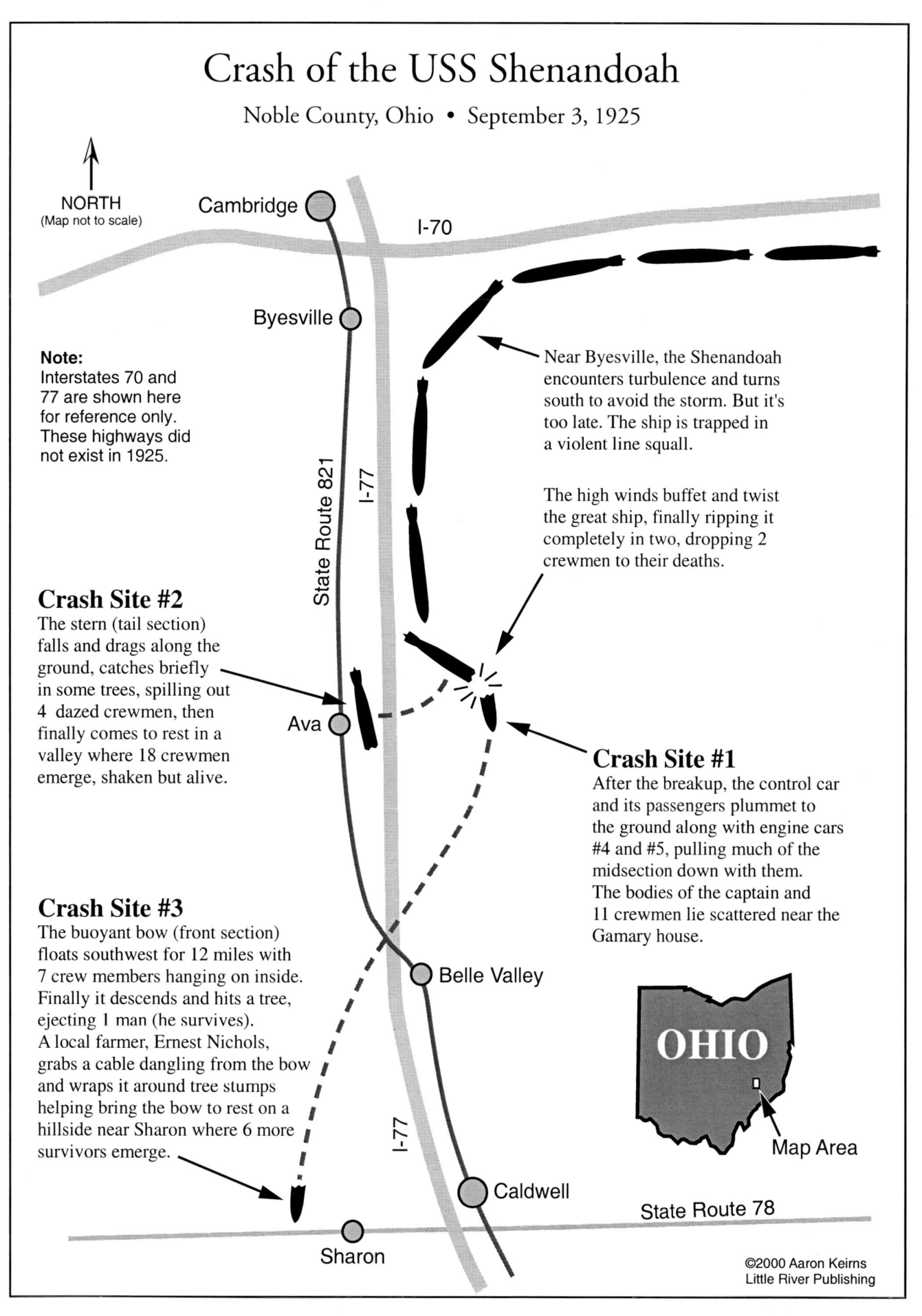

Crash of the USS Shenandoah

Noble County, Ohio • September 3, 1925

NORTH (Map not to scale)

Note: Interstates 70 and 77 are shown here for reference only. These highways did not exist in 1925.

Near Byesville, the Shenandoah encounters turbulence and turns south to avoid the storm. But it's too late. The ship is trapped in a violent line squall.

The high winds buffet and twist the great ship, finally ripping it completely in two, dropping 2 crewmen to their deaths.

Crash Site #2
The stern (tail section) falls and drags along the ground, catches briefly in some trees, spilling out 4 dazed crewmen, then finally comes to rest in a valley where 18 crewmen emerge, shaken but alive.

Crash Site #1
After the breakup, the control car and its passengers plummet to the ground along with engine cars #4 and #5, pulling much of the midsection down with them. The bodies of the captain and 11 crewmen lie scattered near the Gamary house.

Crash Site #3
The buoyant bow (front section) floats southwest for 12 miles with 7 crew members hanging on inside. Finally it descends and hits a tree, ejecting 1 man (he survives). A local farmer, Ernest Nichols, grabs a cable dangling from the bow and wraps it around tree stumps helping bring the bow to rest on a hillside near Sharon where 6 more survivors emerge.

Map Area

©2000 Aaron Keirns
Little River Publishing

Ohio's Airship Disaster 31

Right: This pictorial diagram was part of an article in the January 1925 issue of The National Geographic Magazine.

Below: A poem composed by a local man a few days after the crash.
(Courtesy of Bryan & Theresa Rayner)

Shenandoah

She sailed away
on a September day,
from her moorings
on our eastern coast.

To show in pride,
to a country-side,
her achievements
that we might boast.

O, queen of the air,
in a land so fair,
that you might have
known your fate.

But little you knew
of the winds that blew
o'er the far-famed
Buckeye State.

To the brave who died
on your wild night ride,
our hats, in reverence,
we raise.

To the crew who live,
a salute we give,
with naught but words
of praise.

The yanks of worth
who gave you birth,
we believe are the men
who know.

But how prone we are
to forget the power
of the perilous storms
that blow.

Frank E. Stottlemire
Cambridge, Ohio
September 10, 1925

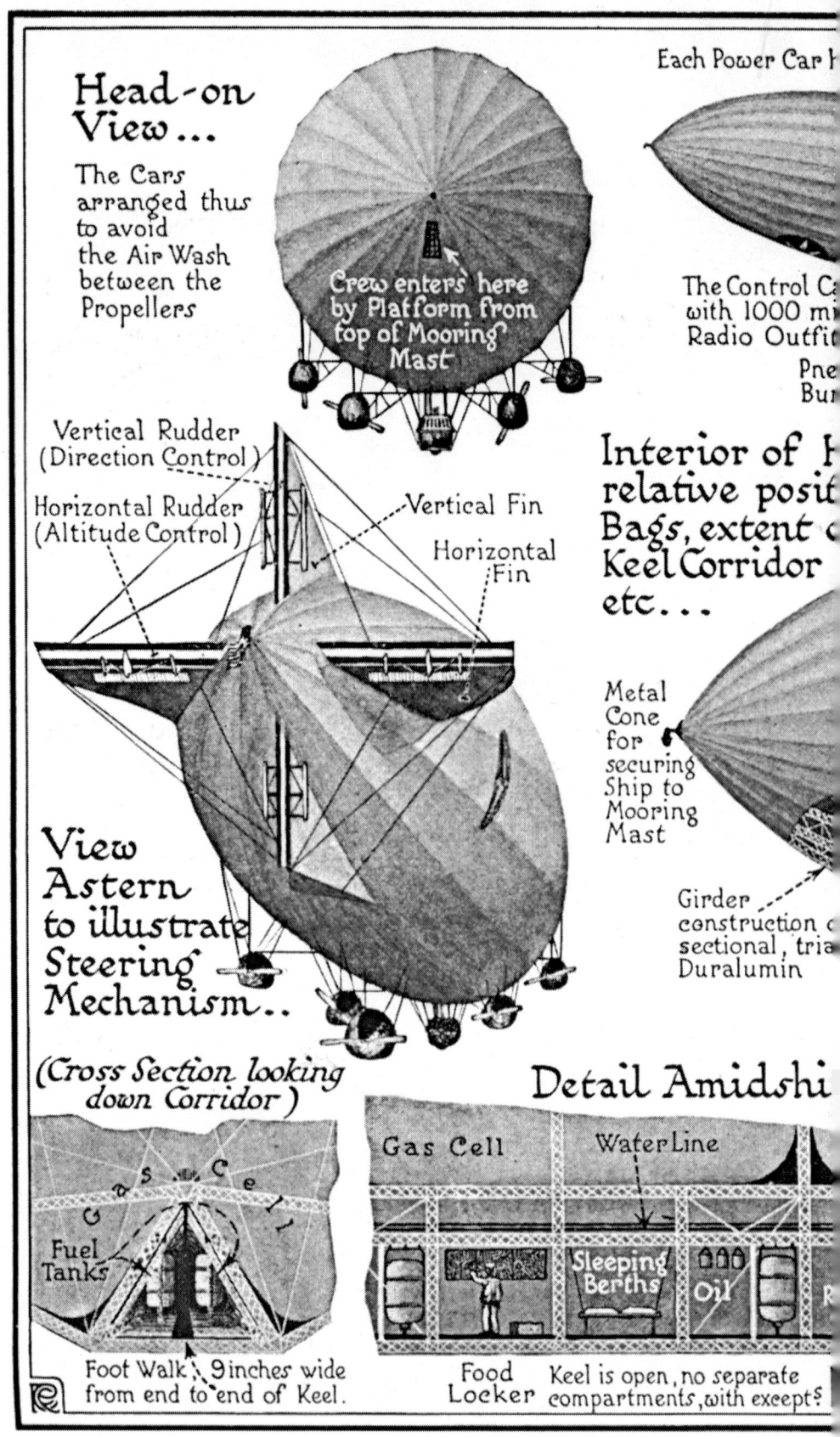

©1925 National Geographic Society – Drawn by Charles E. Riddiford

32   Ohio's Airship Disaster

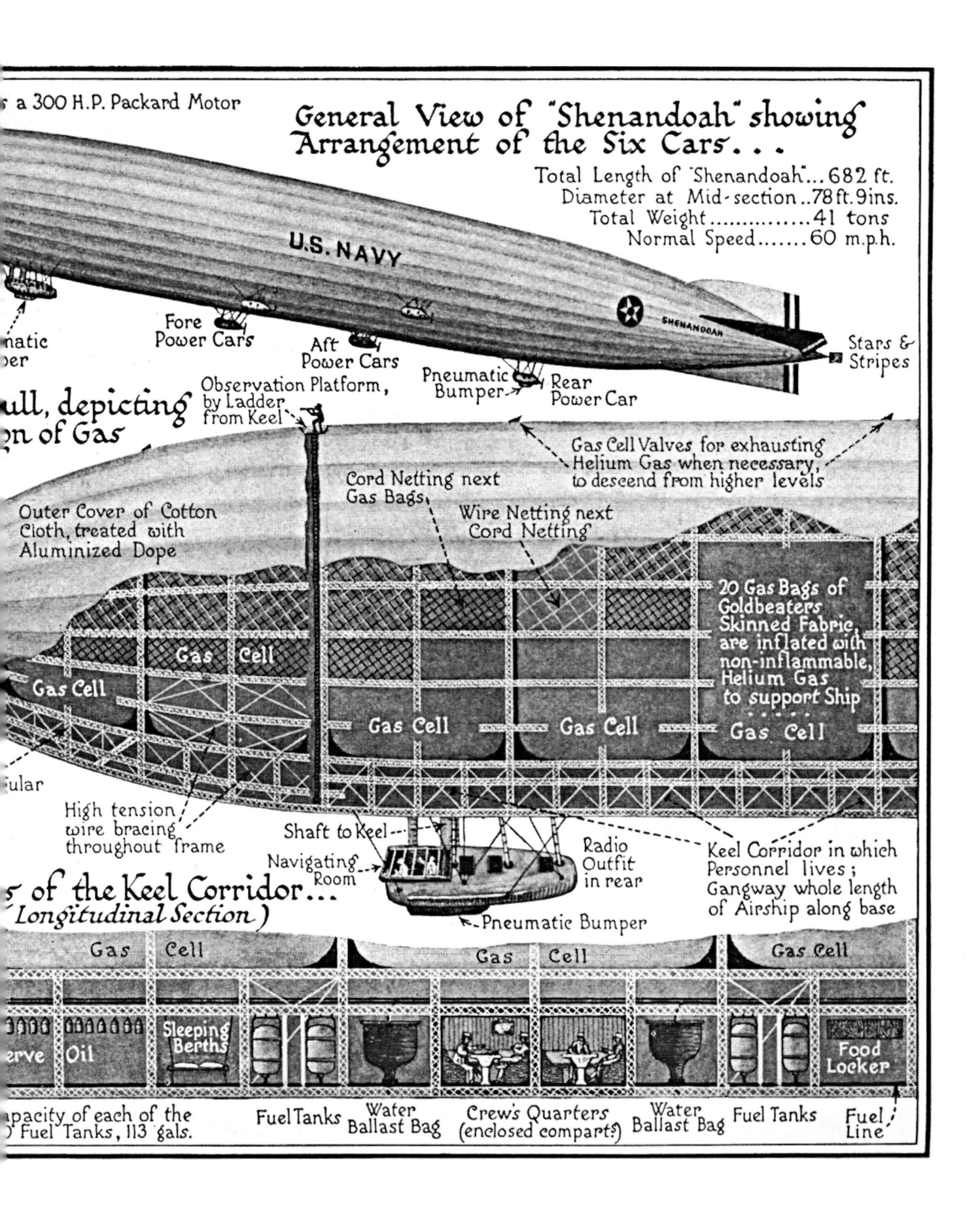

General View of "Shenandoah" showing Arrangement of the Six Cars...

a 300 H.P. Packard Motor

Total Length of "Shenandoah"... 682 ft.
Diameter at Mid-section.. 78 ft. 9 ins.
Total Weight............... 41 tons
Normal Speed....... 60 m.p.h.

U.S. NAVY

Fore Power Cars
Aft Power Cars
Pneumatic Bumper
Rear Power Car
Stars & Stripes

ull, depicting on of Gas

Observation Platform, by Ladder from Keel

Gas Cell Valves for exhausting Helium Gas when necessary, to descend from higher levels

Cord Netting next Gas Bags
Wire Netting next Cord Netting

Outer Cover of Cotton Cloth, treated with Aluminized Dope

20 Gas Bags of Goldbeaters Skinned Fabric are inflated with non-inflammable Helium Gas to support Ship

Gas Cell   Gas Cell   Gas Cell   Gas Cell   Gas Cell

ular

High tension wire bracing throughout frame

Shaft to Keel
Navigating Room
Radio Outfit in rear
Keel Corridor in which Personnel lives; Gangway whole length of Airship along base

Pneumatic Bumper

s of the Keel Corridor...
(Longitudinal Section)

Gas Cell   Gas Cell   Gas Cell

erve Oil Sleeping Berths   Food Locker

apacity of each of the Fuel Tanks, 113 gals.
Fuel Tanks
Water Ballast Bag
Crew's Quarters (enclosed compart's)
Water Ballast Bag
Fuel Tanks
Fuel Line

Ohio's Airship Disaster 33

Above: A view of the Shenandoah's control car or gondola in the hangar at Lakehurst.

Originally the control car of the Shenandoah had an engine at its rear. In order to allow room for the rotation of its 18 ½ ft. propeller, the car was suspended below the hull with struts and cables. Even though the engine was later taken out and replaced with a radio room, the control car remained suspended. This design created a weak point between the car and the hull of the ship which contributed to the car breaking loose during the storm. Later airships were designed with the control car attached directly to the bottom of the ship, eliminating this weak point.

A few days after the crash, Commander Lansdowne was featured on the cover of *Time Magazine*.

THE CRASH CONTINUES

Unlike an airplane crash, the crash of the Shenandoah was a slow process, much like a ship sinking. It took almost a full hour for all the sections and crew members to finally reach the ground.

When the control car broke away from the hull, the front section (bow) suddenly relieved of the weight of the car, began to rise. Soon the severed bow section had risen several thousand feet into the air and was circling with the storm. Inside the bow section, 7 crewmen held on for the ride of their lives.

At the same time, the huge tail section (stern) descended and began to drag and bounce along the ground tail-first toward a little valley not far from the Gamary house. Inside, 22 crewmen were clinging to anything they could hold on to. The stern then caught briefly in some trees ripping off part of the crew section and tossing out 4 of the men. Bruised and dazed, all 4 men survived.

The stern section continued down into the valley where it finally came to rest. All 18 crewmen inside got out alive. Today, this spot is marked by a sign along Interstate 77.

Meanwhile, up in the floating bow section, the 7 occupants drifted with the wind, spinning and traveling in wide circles. After nearly an hour, they managed to release some helium and the 300 ft. bow section began to descend. The bow was almost to the ground when it struck the top of a walnut tree. The branches plucked out one of the crewmen, leaving him dangling unconscious but alive.

As the bow floated over Ernest Nichols' house near the town of Sharon, the men on board shouted down for Ernest to grab one of the dangling trail ropes or mooring wires. Ernest managed to grab a rope and wrap it around a fence post, part of a well and a couple of tree stumps. This helped bring the bow low enough for crewmen to jump out and secure other lines to trees.

The bow was still buoyant however and the sun heating the helium was making it even more so. To keep it from crashing into the farmhouse, crewmen blasted holes in the gas cells with borrowed shotguns.

Finally, nearly an hour after the initial crash, the last section of the Shenandoah had come to rest nearly 12 miles from the initial crash site. All the men on board the floating bow section survived their ordeal.

LIVES LOST

When the tragic crash of the Shenandoah was all over, 14 men had perished, 29 had survived. Most of the crewmen who died were in or near the control car or power cars which broke loose from the hull. Those who happened to be in more buoyant parts of the ship turned out to be the lucky ones. Some of these survivors went on to serve on other airships such as the Los Angeles, Akron and Macon.

As for the dead…at first they were covered up where they lay while officers established their identities and tried to account for all of their

Left top: Shipping crates await the coffins of the fallen airmen in front of the funeral parlor in Belle Valley.
(Pickenpaugh Collection)

Left bottom: A sad sight at the train station in Cambridge.
(Pickenpaugh Collection)

LIST OF CASUALTIES

Caldwell, Ohio, September 3 (A. P.).—The following is a list of the dead in the Shenandoah disaster:

Lieutenant Commander Zachary Landsdowne, Greenville, Ohio.

Lieutenant Commander Louis Hancock, Austin, Texas, executive officer.

Lieutenant J. B. Lawrence, St. Paul, Minn., watch officer.

Lieutenant A. R. Houghton, Allston, Mass., watch officer.

Lieutenant E. W. Sheppard, Washington, D. C., engineer officer.

George C. Schnitzer, Tuckerton, N. J., chief radio man.

James A. Moore, Jr., Savannah, Ga., aviation machinist mate, first class.

Everett P. Allen, Omaha, Neb., aviation chief rigger.

Ralph T. Joffray, St. Louis Mo., aviation rigger.

Bartholomew B. O'Sullivan, Lowell, Mass., aviation machinist mate, first class.

William H. Spratley, Venice, Ill., machinist mate, first class.

Charles H. Broom, Toms River, N. J., aviation machinist mate, first class

Celestino P. Mazzuco, Murray Hill, N. J., aviation machinist mate.

James W. Cullinan, Binghamton, N. J., aviation pilot.

Above: A list of the casualties printed in the Cincinnati Enquirer the day after the crash.

men. Later they were wrapped in blankets and laid on the Gamary's front porch. From there they were taken to the funeral parlor in nearby Belle Valley where 14 simple wooden coffins sat in a row. It took several hours to get authorization from Washington to embalm the bodies. After embalming, the flag-draped coffins were driven north to Cambridge to be put on the train.

Ohio's Airship Disaster 35

36   *Ohio's Airship Disaster*

CHAPTER 5

THE AFTERMATH

News of the Shenandoah crash spread quickly across the Ohio countryside. Soon local folks on the scene were joined by curiosity seekers from surrounding areas. It wasn't long before automobiles full of men, women and children came pouring into the area. They trampled crops and flattened fences.

As people walked around the wreck, they began to pick up little pieces of debris here and there – small momentos of the event. It was little things at first: pieces of metal, fabric, wood and canned goods that had spilled from the ship's food locker.

Then some became more bold and started picking up larger pieces of the metal framework, radio equipment and even personal belongings of the crew. As the crowd grew larger and souvenirs became more scarce, people began twisting off bigger pieces of the wreckage and ripping the silvery fabric from the hull. Some folks left with armloads of fabric and long pieces of metal framework tied to their running boards.

One enterprising tailor quickly began producing and selling "Shenandoah Slickers" (raincoats) made from large sections of material taken from the helium cells *(see sidebar this page)*.

The surviving officers and enlisted men tried to police the souvenir hunters, but with little success. There were just too many people.

It wasn't long before the fallen stern and bow of the airship had been picked clean. Any fabric that could be reached had been torn off, any loose metal had been carried away. Some folks even used hammers and wrenches in their quest for souvenirs. Soon the once-proud Daughter of the Stars had been reduced to a mere skeleton.

Local stores began to set up displays in their front windows with pieces and artifacts from the fallen ship. As the day wore on, the atmosphere at the crash sites became almost festive. There seemed to be a sense that the ship belonged to no one – and everyone. Farmers began charging people to drive onto their land, ostensibly in an effort to recoup their losses from trampled crops and fences.

The National Guard was finally called in to take control.

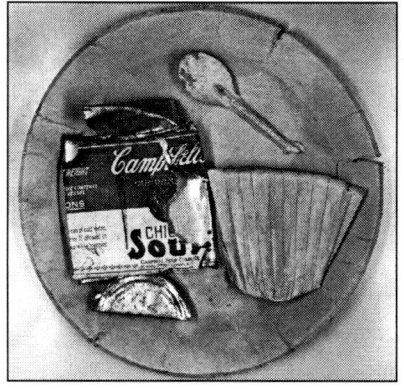

Above: A can of soup and eating utensils picked up at the crash site. To keep the airship as light as possible, plates and spoons were made of pressed paper.
(Pickenpaugh Collection)

Opposite page: The remains of the stern section.
(Pickenpaugh Collection)

Helium Cells

The large bags or cells containing the helium gas were made of a material called "goldbeaters skin." The material consisted of linen lined with layers of membrane from the intestines of cattle. Goldbeaters skin was gas-tight, flexible and rip resistant.

Artisans often used this material to contain gold as they hammered in into gold leaf – hence the name "goldbeaters skin."

38 *Ohio's Airship Disaster*

"The crowds of curious ate Andy out of house and home, drank his well dry, tramped out his garden, and killed most of the grass on his little place."
(Caldwell Press, 12-30-25)

Crash Site #1

The control car, engine cars #4 and #5, and much of the midsection fell near the Gamary house. The crew space became entangled in the trees down over the hill.

Left: A tattered photograph showing a view of the wreckage at Andy Gamary's house near Ava.
(Pickenpaugh Collection)

Below: The same view of the old Gamary house as it looks today.
(A. Keirns Photo)

Ohio's Airship Disaster 39

Garden of T. R. Davis, near Ava, O. where 13 men died in wreck of SHENANDOAH. Sept 3, '25

Above: Another view of the wreckage at Andy Gamary's house near Ava. The Gamarys were renters, the house was actually owned by T. R. Davis.
(Brockway Collection)

Right: A blurry aerial view of the Gamary house and surrounding area. Numbers written in white point out specific areas of the crash scene. #1 points to where the control car came down, #2 where two engine cars and part of the midsection fell, and #3 (at the top of the photo) where the crew space became entangled in the trees.
(Pickenpaugh Collection)

40   *Ohio's Airship Disaster*

Left: The granite marker at Site #1 near the Gamary house.
(A. Keirns Photo)

Left: A sandstone marker, carved by a local man, marks the spot where Commander Lansdowne fell.
(A. Keirns Photo)

Ohio's Airship Disaster 41

Aft engine car #1 landed down over the hill from the Gamary house, about 100 ft. from the front end of the stern section.

Right & Below: Three views of engine car #1. Notice the handrail used by the ground crew to help maneuver the airship during launchings and landings (as shown in the small photo below).

(Pickenpaugh Collection)

42   *Ohio's Airship Disaster*

A soldier guards engine car #1.
(Pickenpaugh Collection)

44   *Ohio's Airship Disaster*

Crash Site #2

The stern section, along with engine cars #2 and #3, came to rest on the Neiswonger farm down over the hill from the Gamary house.

Ohio's Airship Disaster

46   *Ohio's Airship Disaster*

Opposite Page & Above: Views of the stern section at crash site #2.
(Pickenpaugh Collection)

Left: The wreckage of engine car #3, near the aft of the stern section.
(Brockway Collection)

Below: Today site #2 is marked by this sign along Interstate 77.
(A. Keirns Photo)

Ohio's Airship Disaster   47

Above: The stern section after much of the outer fabric had been torn off and carried away.
(Brockway Collection)

Right: A group of nurses stand near the wreckage of engine car #3.
(Brockway Collection)

Opposite page: The crumpled tail of the airship. Notice the soldier (near center of picture) patrolling inside the roped-off area.
(Brockway Collection)

48   Ohio's Airship Disaster

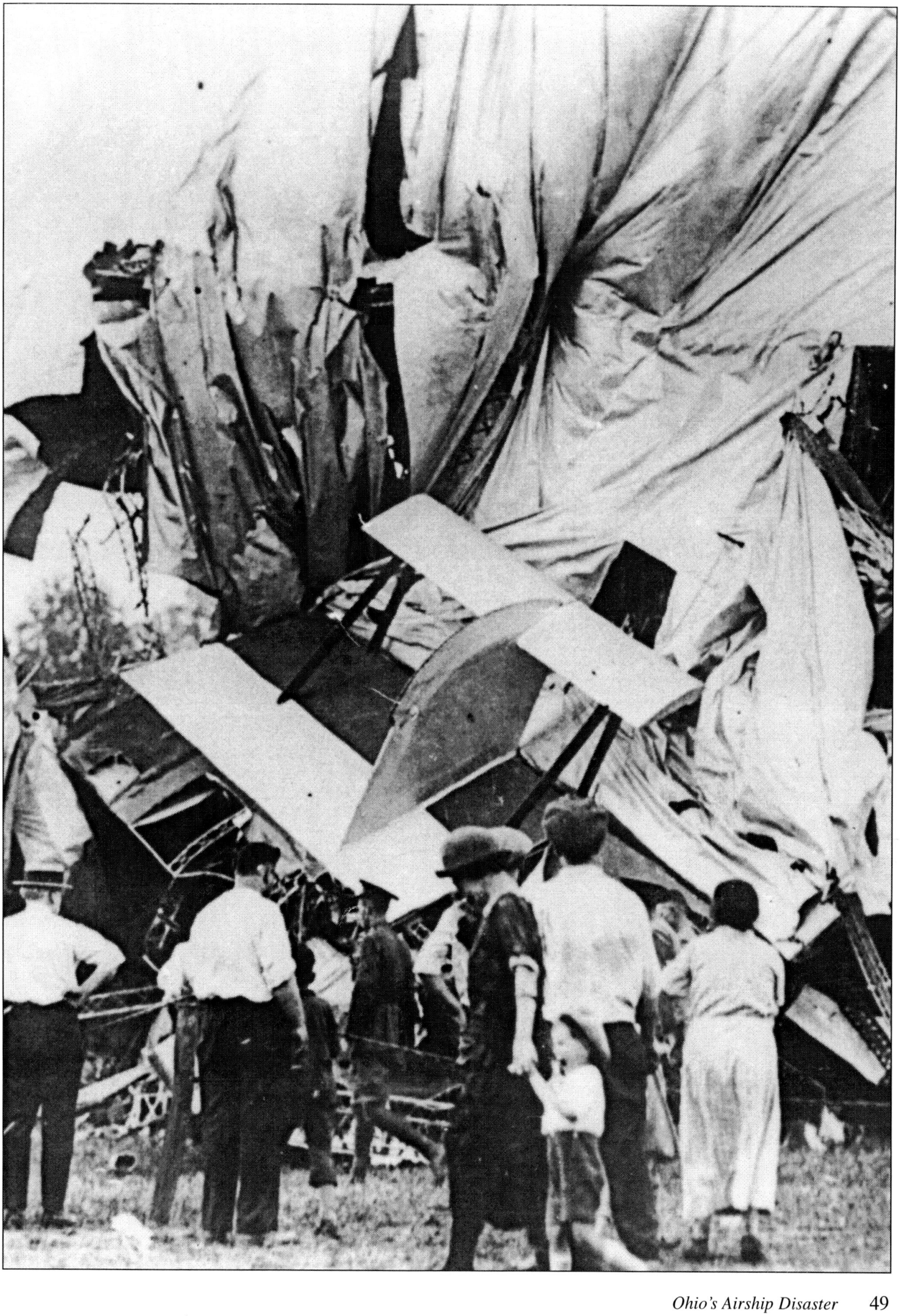

Front Section Shenandoah Disaster Sharon Ohio Sept 3.

50 *Ohio's Airship Disaster*

Crash Site #3

The bow section landed on the farm of Ernest Nichols about 3 miles west of Caldwell near the town of Sharon.

Above: Panoramic view of the bow section after most of the outer fabric had been torn away.
(Clements Photo, Keirns Collection)

Opposite page: The nose of the airship. Notice the mechanism in the center of the nose cone used in attaching the ship to mooring masts.
(Pickenpaugh Collection)

Left & Below: Markers at Site #3 along State Route 78 near Sharon.
(A. Keirns Photos)

Ohio's Airship Disaster   51

52   *Ohio's Airship Disaster*

CHAPTER 6

THE LEGACY OF THE SHENANDOAH

Three quarters of a century has now passed since the Shenandoah dropped out of the sky into the remote hills of Noble County, Ohio. All we have left to remember her by are a few scraps of metal and fabric, old pictures and memories.

In 1937 a monument was erected in Ava to pay honor to those who perished in the crash. In 1975, on the 50th anniversary of the crash, a ceremony was held to commemorate the crash. Local dignitaries, radio personality Paul Harvey, and many others attended. Even former Rigger John McCarthy was there. John was the fellow who was so violently plucked out of the floating bow section by the branches of a tree on that September day 50 years earlier.

Since the 50th anniversary, interest in the Shenandoah crash has been kept alive mainly by a few local folks. Probably the most notable of these has been the Rayner family of Ava. Rayner's Garage at the edge of Ava has become the unofficial information center for the Shenandoah crash.

Bryan and Theresa Rayner have had a life-long interest in the Shenandoah. Both of Bryan's grandfathers were at the scene soon after the crash. His maternal grandfather sold pop to the thirsty crowds. For years afterward, Bryan's paternal grandfather gave tours of the crash sites and young Bryan often tagged along. In 1995 Bryan and Theresa transformed a little camping trailer into a mobile Shenandoah Museum. In it they placed artifacts, pictures and other memorabilia related to the crash. They often pull their museum around to schools and local events to help keep the memory of the Shenandoah alive. In 1998 a disastrous flood hit the Ava area. The Rayner's museum suffered serious water damage. Undeterred, they cleaned it up, dried it out and put it all back together. We are lucky to have these devoted guardians of Shenandoah history.

HUMAN NATURE

Much has been written and said about the frenzy of looting or souvenir-hunting that took place when crowds converged on the crash scenes. There are, no doubt, psychological or sociological theories that

Above: A 20¢ postage stamp cancelled with a commemorative postmark at the 70th anniversary of the Shenandoah crash.

Opposite page: A variety of memorial covers commemorating the crash of the Shenandoah and other airships.

Above: This piece of frame and metal sink from the Shenandoah are on display in the Rayners' mobile museum. *(A. Keirns Photo)*

may explain why these folks acted the way they did. After all, these were not gangs of wild outlaws or invading barbarians. They were just ordinary people – caught up in an extraordinary situation. This was not the first time something like this had happened, nor would it be the last.

There was one facet of the event that was relatively new however. The advent of radio allowed the news of the crash to be broadcast almost immediately to thousands of people at the same time. This was a new phenomenon and it changed things. In the past, news had traveled much slower especially in rural areas. But by 1925 radio was there to send out the message and many folks now had cars (or knew someone who did) that would take them there quickly. Instead of waiting to read about the crash in the next day's (or next week's) newspaper, they could go to the scene and experience it firsthand.

A similar scenario had occurred earlier that same year in Kentucky. An experienced cave explorer named Floyd Collins became wedged in a small passageway deep in Sand Cave. For several days rescuers tried to save the unfortunate Floyd. Radio stations began broadcasting live reports and regular updates on the situation. They were joined by a swarm of newspaper reporters and film crews in what may have been the world's first "media circus." Curiosity seekers rolled in by the carload – thousands of them. Although there weren't many souvenirs to scavenge in this case, the atmosphere at the scene developed into an air of festiveness, much like at the Shenandoah crash scene.

Folks set up stands to sell food and drink at ridiculously high prices to the curiosity seekers. Hawkers sold balloons and other trinkets. Jugglers even performed. Like at the Shenandoah crash, the National Guard finally had to be called in to keep order.

Meanwhile, deep below the feet of these curiosity seekers, poor Floyd was fighting for his life. Despite a variety of rescue attempts, Floyd was dead by the time they finally reached him. The story was over. The media circus packed up and went home.

In the case of the Shenandoah crash, many folks had been attending the Noble County Fair in nearby Caldwell when they heard about the fallen airship. They were already dressed-up and in a festive mood before they even got there. They simply left one crowded attraction to go see another.

Morbid Curiosity

Both Floyd Collins and the Shenandoah were victims of morbid curiosity – one of human nature's more bizarre personality quirks. There seems to be something deep in the primitive part of our brains that attracts us to the very things that repulse and frighten us. Death, for example.

When the Shenandoah fell out of the sky in rural Ohio, folks didn't rush to the scene to plunder the fallen airship. They came because it was the biggest thing that had ever happened in their rural community. A giant, exotic flying machine had tumbled to earth out of a turbulent cloud. Many of them didn't know exactly what this bizarre thing was. Some, like Andy Gamary, had never even heard of the Shenandoah.

For others, the Shenandoah was a celebrity. They had read about her

Above: The Shenandoah monument located in Ava, Ohio.

Left: Detail of the bronze sculpture near the top of the monument.
(A. Keirns Photos)

Ohio's Airship Disaster 55

Opposite page top: Bryan Rayner holding a model of the Shenandoah outside his mobile museum.

Opposite page bottom: The inside of the Rayners' Shenandoah museum.
(A. Keirns Photos)

Above: A glass container from the airship still holding its contents of sugar cubes, is on display in the Rayners' mobile museum.
(A. Keirns Photo)

Ring Bearer

When Mrs. Larrison found Zachary Lansdowne's class ring in her garden, she called the county sheriff. The sheriff notified the FBI. It wasn't long before two FBI agents showed up in Noble County. They carried the ring back to FBI headquarters where the director of the FBI, J. Edgar Hoover, took a personal interest.

When the long-lost ring was finally delivered to Zachary Lansdowne's widow, it was by none other than J. Edgar himself.

---

transcontinental trip and just seeing her fly over would be an exciting and memorable event. Now, America's famed Daughter of the Stars had fallen into their back yards! It was huge, literally, and they had to see it.

These folks weren't insensitive to the fact that lives had been lost in the crash. They didn't come to desecrate the dead – they came out of natural curiosity. But, as history has proven many times, people do things as part of a crowd that they would never do alone. It's human nature to want to take away a souvenir from such a once-in-a-lifetime event. But as this simple act of picking up a souvenir became multiplied by thousands of hands, it swelled into a destructive force.

It can't be ignored that there were those who gathered up as much of the fallen airship as they could with the idea of making money from the disaster. We'd like to think that these opportunists represented only a small fraction of those who visited the crash sites. Noble Countians are quick to point out that much of the looting was done by outsiders.

It's been 75 years since this all happened. Who can say why the crowds acted the way they did on that day so long ago. It's useless to condemn or assign blame at this late date. Right or wrong, it was human nature. Somehow, amidst the chaos of it all, it seems the tragedy of the event became overshadowed by the magnitude of the spectacle. That's about all we can say for sure.

THE MYSTERY OF THE RING

One of the more curious stories to come out of the crash was the story of the lost ring. When the control car crashed into the Gamary garden, Lieutenant Commander Lansdowne was killed instantly. But when his body was taken to the funeral parlor to be embalmed it was noticed that his Annapolis class ring was missing from his finger. A further search of the area where he fell produced nothing. It was assumed that the ring was either taken by an unscrupulous souvenir-hunter or lost somewhere in the chaos of the crash.

Years go by. Then in early June of 1937, a Mrs. Larrison is doing some weeding in the old Gamary garden when something shiny catches her eye. There, impaled on a mustard plant, is the missing ring.

Soon the ring is reunited with the Lansdowne family. After nearly 12 years it seems the earth had given up its prize. But some were skeptical of the find. The commander's daughter, Peggy, remembers her father having trouble with his ring. The onyx was prone to cracking and would easily bump loose. Yet, after all the ring had been through, the onyx of the returned ring was in perfect condition.

Maybe the commander had his ring fixed before the flight, or maybe the ring was cleaned and repaired after it was found. Who knows.

A coincidental, but apparently overlooked, fact about this story involves the particular time at which the ring chose to reappear. If indeed the ring was found in early June of 1937, it would have been only 3 or 4 weeks after the terrible crash of the airship Hindenburg took 36 lives. This crash, with its graphic photographs, film reels and emotional radio broad-

Ohio's Airship Disaster   57

Right: The returned ring, now displayed in an engraved case.
(C. Lansdowne Hunt)

Opposite page top: Cover of the sheet music for the Wreck of the Shenandoah song.

Opposite page bottom: A player piano roll of the same tune.
(Rayner Collection)

Far Right: The words of the Wreck of the Shenandoah as written by Maggie Andrews. Interestingly, Vernon Dalhart's recorded rendition of the song didn't include the third and fourth stanzas.

Below: The 36" scale model of the Shenandoah on display in the Rayners' mobile museum.
(A. Keirns Photo)

cast, no doubt stirred-up many memories and emotions for folks who had been at the site of the Shenandoah crash. For whatever reason, maybe it was time for the ring to go home.

A Song to Remember Her By

During the era when the Shenandoah crashed it was common to commemorate tragic events with a song. One of the most prolific country singers of this period was Vernon Dalhart. In 1925 Dalhart recorded the *Wreck of the Shenandoah*, a song written by Maggie Andrews.

We'll end our story of the Shenandoah with the words to this sentimental song (see following page). Its simple verses cut through the clutter of the crash to lament the human tragedy of lives lost.

58   *Ohio's Airship Disaster*

The Wreck of the Shenandoah

Words & Music by Maggie Andrews
©1925 by Shapiro Bernstein & Co., Inc.

At four o-clock one evening
Of a warm September day
A great and mighty airship
From Lakehurst flew away

The mighty Shenandoah
The pride of all this land
Her crew was of the bravest
Captain Lansdowne in command

The giant motors thundered
She proudly sailed along
Each man was at his station
Each heart was true and strong

They started for St. Louis
As day turned into night
With not a thought of danger
On that sad and fatal flight

At four o-clock next morning
The earth was far below
Then a storm in all its fury
Gave her a fatal blow

For hours they bravely struggled
They worked with all their might
But the storm could not be conquered
And the ship gave up the fight

Her sides were torn asunder
Her cabin was torn down
The captain with his brave men
Went crashing to the ground

And fourteen lives were taken
But they've not died in vain
Their names will live forever
Within the Hall of Fame

In the little town of Greenville
A mother's watchful eye
Was waiting for the airship
To see her son go by

But alas her boy lay sleeping
His last great flight was o'er
He's gone to meet his Maker
His ship will fly no more

A loving wife and children
A mother's broken heart
They're mourning for their loved one
Since the storm tore them apart

But their faith will not be shaken
They'll see him bye and bye
They know he waits in Heaven
Where the brave go when they die

Ohio's Airship Disaster

Addendum

America's involvement with lighter-than-air flight didn't end with the Shenandoah.

The Los Angeles continued flying and was finally decommissioned in 1932.

After the Shenandoah there were two more American-made dirigibles – the Akron (ZR-4) and the Macon (ZR-5).

In 1933 the Akron crashed in a storm over the Atlantic, killing 73 of the 76 men on board. In 1935 the Macon crashed into the Pacific, killing 2 of the 83 men on board.

As the Macon sank beneath the waves, so ended America's rigid dirigible program.

Above: The obverse and reverse of a silver medal commemorating the Akron and Macon, minted around the early 1970's.

Glossary of Terms

Airship - A lighter-than-air craft having propulsion and steering systems.

Ballast - A heavy substance (usually water in airships) used to improve the stability and control of the ascent and descent of a balloon or airship. For example, if the ship begins to descend too quickly, ballast can be released to slow the fall.

Blimp - An airship lacking a rigid internal support structure.

Bow - The forward part of a ship or airship.

Buoyant - Capable of floating.

Cell - In airships, a bag containing the buoyant gas.

Control Car - The compartment from which the airship is controlled.

Dirigible - An airship which is capable of being steered.

Duralumin - A light, but strong alloy of aluminum, copper, manganese and magnesium used in the framework of airships such as the Shenandoah.

Elevator - A moveable airfoil or wing attached to the tail of an airship to control direction of flight in a vertical plane (up and down).

Flammable - Capable of easily igniting and burning.

Goldbeaters Skin - The material making up the gas cells of an airship, made of linen lined with layers of membrane from the intestines of cattle.

Gondola - An elongated car attached to the underside of an airship.

Helium - A light, colorless, non-flammable gas used in lighter-than-air craft.

Hot Air Balloon - A lighter-than-air craft whose lift is provided by heated air, usually having no propulsion or steering systems.

Hull - The frame or body of a ship or airship.

Hydrogen - A light, colorless, highly-flammable gas used in early lighter-than-air craft.

Inert Gas - A gas that exhibits great stability and low reaction rates.

Keel - The structural supporting member of a ship or airship that extends longitudinally along the center of its bottom.

Line Squall - A squall or thunderstorm occurring along a cold front.

Mooring - In airships, a tall mast to which the nose of the airship is secured while not in flight.

Pneumatic Bumpers - On the Shenandoah, bumpers, controlled by compressed air, located on the underside of the control car and rear engine car to help prevent damage should the airship contact the ground during launching or landing.

Rigid Airship - An airship whose shape is defined by a rigid internal structure. The lifting gas is contained in bags or cells within the rigid framework.

Rudder - A moveable airfoil or wing attached to the tail of an airship to control direction of flight in a horizontal plane (side to side).

Semi-rigid Airship - A blimp with a rigid keel.

Stern - The rear end of a boat or airship.

Strut - A structural member designed to provide strength in the direction of its length. On the Shenandoah, the control car and engine cars were attached to the hull of the ship with struts.

Zeppelin - A rigid airship.

60   *Ohio's Airship Disaster*

ACKNOWLEDGMENTS

*N*obody writes a book like this without help. It takes a lot information and encouragement from a variety of people to make it happen.

My cousin, Bill Keirns, and his wife Barb were the ones who first told me about the 70th anniversary event that was being held in Ava in 1995. I attended that event, and I was hooked. I wanted to learn more about this unusual slice of Ohio history.

Later I met Bryan and Theresa Rayner of Ava who have become my local contacts in the area and a great source of information. Bryan has shown me around the crash sites and they are both always eager to help in any way they can.

At the 70th anniversary event I briefly met C. Lansdowne Hunt, the grandson of the Shenandoah's commander, Zachary Lansdowne. He gave a touching speech to the crowd assembled at the Shenandoah monument that day. Five years later I asked him to write the foreword for this book and he graciously agreed.

Much of the story of the Shenandoah is told with pictures. Again, Bryan Rayner helped me in that area along with Fern Pickenpaugh and my old friend Chance Brockway of Buckeye Lake, Ohio.

And, without the constant support of my wife, Bernice, and our family, I wouldn't be able to pursue these book-writing adventures.

A. K.

ABOUT THE AUTHOR

Aaron Keirns is a native Ohioan with a life-long interest in the history of the Buckeye State.

He is also the author of the popular books: *Black Hand Gorge...A Journey Through Time*, and *Statues on the Hill*.

Aaron holds a degree in Anthropology and is a writer, publisher, award-winning graphic designer and entertaining speaker.

He and his wife, Bernice, live near Howard, Ohio. They have four children: Adam, Tracy, Jesse and Nathan – and three grandchildren: Elisa, Siaira and Ethan.

Bibliography

Books:

BOTTING, Douglas "The Giant Airships,"
(The Epic of Flight; 6) Time-Life Books, Alexandria, 1981.

COOKE, David C. "Dirigibles That Made History,"
G. P. Putnam's Sons, New York, 1962.

DEIGHTON and SCHWARTZMAN "Airshipwreck,"
Holt, Rinehart and Winston, New York, 1978.

GRAY, Lewis H. "Ill Wind,"
Gateway Press, Inc., Baltimore, 1989.

HOOK, Thom "Shenandoah Saga,"
Air Show Publishers, Annapolis, 1973.

JABLONSKI, Edward "Man With Wings,"
Doubleday & Company, Inc., New York, 1980.

RICHARDS, Norman "Giants in the Sky,"
Children's Press, Chicago, 1967.

TOLAND, John "Ships in the Sky,"
Henry Holt & Co., New York, 1957.

Articles:

DICHRISTINA, Mariette "What Really Downed the Hindenburg,"
Popular Science, November, 1997.

NELSON, Larry L. "Disaster at Dawn,"
Timeline, Dec./Jan., 1991.

PICKENPAUGH, Roger "The Day the Daughter of Stars Fell to Earth,"
Country Living, (a publication of Ohio Rural Electric Cooperatives) Sept., 1995.

TALMAN, Charles Fitzhugh "Who's Who Among the Storms,"
Nature Magazine, November, 1925.

TOLAND, John "Death of a Dirigible,"
American Heritage, Feb., 1969.

WOOD, Junius B. "Seeing America from the Shenandoah,"
The National Geographic Magazine, January, 1925.

Zachary Lansdowne
1889 – 1925
Captain of the USS Shenandoah

In the little town of Greenville
A mother's watchful eye
Was waiting for the airship
To see her son go by

But alas her boy lay sleeping
His last great flight was o'er
He's gone to meet his Maker
His ship will fly no more

Excerpted from: *The Wreck of the Shenandoah*
by Maggie Andrews

Ohio's Airship Disaster 63